Backyard Scientist to Lab Pioneer: How to Forge a Path in Biotechnology Research

By Silas Meadowlark

Index

- Discovering Your Passion for Biotechnology
    - Exploring the Wonders of the Living World
    - Understanding the Impact of Biotechnology
    - Identifying Your Unique Interests and Strengths

- Laying the Groundwork: Educational Pathways
    - Navigating the Education System
    - Choosing the Right Degree Program
    - Developing a Solid Foundation in Science

- Building a Backyard Lab
    - Gathering Essential Equipment and Supplies
    - Conducting Experiments at Home
    - Mastering Lab Techniques and Safety

- Networking and Mentorship
    - Connecting with Professionals in the Field
    - Seeking Out Experienced Mentors
    - Leveraging Internships and Volunteer Opportunities

- Advancing Your Skills: Hands-on Experience
    - Participating in Research Projects
    - Attending Workshops and Conferences
    - Honing Your Critical Thinking and Problem-Solving Abilities

- Navigating the Academic World
    - Applying to Graduate Programs
    - Excelling in Research and Coursework
    - Balancing Academia and Personal Life

- Embracing the Entrepreneurial Spirit
    - Identifying Market Opportunities
    - Developing Innovative Products or Services

- Crafting a Successful Business Plan
- Funding Your Biotechnology Endeavors
  - Exploring Research Grants and Fellowships
  - Crowdsourcing and Crowdfunding Strategies
  - Navigating the World of Venture Capital
- Overcoming Challenges and Setbacks
  - Developing Resilience and Perseverance
  - Learning from Failures and Mistakes
  - Maintaining a Positive Mindset
- Navigating the Regulatory Landscape
  - Understanding Biotechnology Laws and Regulations
  - Ensuring Ethical and Responsible Practices
  - Collaborating with Regulatory Agencies
- Translating Research into Real-world Applications
  - Identifying Practical Solutions to Global Challenges
  - Bridging the Gap Between Lab and Industry
  - Communicating Complex Concepts to Diverse Audiences
- Embracing Diversity and Inclusion
  - Fostering Inclusive Environments
  - Empowering Underrepresented Groups in Biotechnology
  - Advocating for Equity and Representation
- Achieving Work-Life Balance
  - Prioritizing Self-care and Well-being
  - Managing Time and Stress Effectively
  - Cultivating a Supportive Network
- Leaving a Lasting Legacy
  - Mentoring the Next Generation of Biotechnology Pioneers
  - Contributing to the Advancement of the Field
  - Inspiring Others through your Achievements
- The Future of Biotechnology: Trends and Innovations
  - Emerging Technologies and their Implications

- Addressing Global Challenges with Biotechnology
- Envisioning the Next Frontier of Scientific Discovery

Discovering Your Passion for Biotechnology

Exploring the Wonders of the Living World

Prepare to start on a thrilling journey into the captivating realm of biotechnology, where the very fabric of life unfolds before your curious gaze. From the detailed dance of DNA to the breathtaking symphony of cellular processes, the living world is a veritable wealth of mysteries waiting to be figured out. Immerse yourself in the awe inspiring beauty of nature's engineering, where even the smallest microbes possess a convolution that puts the most advanced human made machines to shame. As you explore deeper into this field, you'll find yourself constantly amazed by the sheer elegance and efficiency of biological systems, and you'll be inspired to uncover their secrets and employ their power for the betterment of humanity.

Understanding the Impact of Biotechnology

Biotechnology is more than just a scientific discipline; it is a radical force that has the potential to reshape the very fabric of our world. From revolutionizing healthcare by developing trailblazing treatments and therapies, to revolutionizing agriculture by improving crop yields and sustainability, the applications of biotechnology are as diverse as they are deep. Imagine a future where we can engineer bacteria to clean up environmental pollutants, or where we can design personalized cellular therapies to eradicate devastating diseases. The possibilities are truly mind bending, and as a budding biotechnology pioneer, you have the opportunity to be at the forefront of these revolutionary advancements. Embrace the power of this field and let your imagination soar, for the impact you can make is limited only by the boundless potential of the living world.

Identifying Your Unique Interests and Strengths

As you explore into the captivating world of biotechnology, it's important to take the time to explore and identify your unique interests and strengths. Are you fascinated by the detailed workings of the human genome and the potential of gene editing? Do you thrive on the challenge of developing novel biofuels or finding original solutions to environmental challenges? Or perhaps you're drawn to the medical applications of biotechnology, where you can play a crucial role in improving human health and wellbeing. Whatever your passions may be, it's essential to find the perfect intersection between your interests, your aptitudes, and the ever evolving needs of the biotechnology terrain. Embrace your individuality, trust your instincts, and let your curiosity guide you on this exciting journey. Remember, the biotechnology field is vast and diverse, offering ample opportunities for you to carve out a niche that syncs with your unique talents and aspirations. Begin on this odyssey with an open mind, and you'll be well on your way to becoming a trailblazer in the captivating world of biotechnology.

Laying the Groundwork: Educational Pathways

Navigating the Education System

Starting on your biotechnology journey can be a thrilling yet daunting endeavor, with a many of educational paths to consider. Fear not, my fellow aspiring lab pioneers - navigating the education system is akin to weaving through a labyrinth of possibilities, each turn leading you closer to your scientific destiny. The key is to approach this journey with a curious mindset, embracing the twists and turns as opportunities to expand your horizons.

Begin by immersing yourself in the diverse array of academic programs available, from traditional university degrees to specialized biotechnology focused curricula. Scour the course catalogs, attend virtual open houses, and engage with current students to gain a fine understanding of the strengths and unique selections of each institution. Remember, the perfect fit for your aspirations may not be the one with the flashiest reputation or the highest rankings - it's about finding the institution that harmonizes with your passions and authorizes you to thrive.

As you navigate this educational labyrinth, be prepared to navigate the ever evolving setting of funding opportunities. Scholarships, grants, and research assistantships can be your ticket to making accessible the doors to your biotechnology dreams. Embrace your inner sleuth and scour the depths of the internet, attend informational sessions, and network with faculty members to uncover the hidden gems that can propel your academic journey forward.

Ultimately, the path to becoming a biotechnology pioneer is not a one size-fits all endeavor. It's a choose your-own adventure story, where your curiosity, resilience, and unwavering dedication will be your guiding lights. Embrace the challenges, celebrate the small victories, and trust that every step you take will bring you closer to realizing your aspirations of shaping the future of this dynamic field.

Choosing the Right Degree Program

Navigating the vast array of degree programs in the field of biotechnology can be akin to peering into a kaleidoscope of possibilities, each option shimmering with its own unique blend of disciplines and specializations. As you start on this journey, it's vital to approach the decision making process with a keen eye for detail and a unceasing pursuit of finding the perfect fit.

Begin by delving into the diverse array of undergraduate programs, from traditional biology and chemistry degrees to more specialized tracks in areas like bioengineering, bioinformatics, or even entrepreneurship focused biotechnology programs. Each of these paths offers a unique lens through which to view the vast setting of the living world, and the key is to identify the one that connects most deeply with your passions and strengths.

As you progress through your academic journey, the decision making process becomes even more complicated, with a plenty of graduate level programs vying for your attention. From research focused master's degrees to cutting edge Ph.D. programs, the options can seem dizzying. But fear not, my fellow biotechnology enthusiasts - this is where your newfound expertise and detecting eye will shine.

Immerse yourself in the research areas and faculty specialties of each program, seeking out the ones that coordinate most closely with your own scientific interests and aspirations. Engage in virtual information sessions, connect with current students, and even reach out to potential mentors to gain a deeper understanding of the program's unique culture and opportunities. Remember, the right degree program is not just about the letters after your name - it's about finding the environment that will nurture your growth, challenge your thinking, and propel you towards your ultimate goals.

As you navigate this labyrinth of educational options, trust your instincts, embrace your curiosity, and never lose sight of the greater purpose that drives you - to become a pioneering force in the ever evolving realm of biotechnology. The path may not be linear, but with each step, you'll inch closer to uncovering the secrets of the living world and leaving an indelible mark on the future of science.

Developing a Solid Foundation in Science

In the captivating world of biotechnology, a solid foundation in the sciences is akin to a sturdy foundation upon which you can build your dreams of becoming a laboratory trailblazer. It's the bedrock that will support your ascent into the realms of cutting edge research, inventive product development, and radical discoveries.

Begin by immersing yourself in the fundamental disciplines that form the backbone of biotechnology - biology, chemistry, and mathematics. These core subjects will equip you with the essential tools and conceptual frameworks necessary to navigate the complex and ever evolving setting of the living world.

Embrace the challenge of mastering these subjects, for they will serve as the springboard from which you can launch your forays into more specialized biotechnology domains.

As you explore deeper into your studies, seek out opportunities to engage in hands on laboratory experiences. From conducting experiments in your own backyard setup to participating in cutting edge research projects at renowned institutions, these practical experiences will not only solidify your understanding of scientific principles but also ignite your passion for the art of inquiry and discovery.

Embrace the power of interdisciplinary collaboration, for the future of biotechnology lies in the intersections of diverse fields. Forge connections with peers from computer science, engineering, and even the social sciences, as their unique perspectives and skill sets can broaden your horizons and inspire new solutions to complex problems.

Remember, the journey of developing a solid foundation in science is not a linear path, but rather a dynamic and ever evolving process. Stay curious, embrace failure as an opportunity for growth, and never lose sight of the bigger picture - the power of biotechnology to transform our world and improve the human condition. With each passing semester, each challenging lab session, and each moment of breakthrough, you'll find yourself one step closer to becoming a true biotechnology pioneer.

Building a Backyard Lab

Gathering Essential Equipment and Supplies

Begin on your biotechnology journey by transforming your backyard into a bustling hub of scientific exploration. The first step? Gathering the essential equipment and supplies to build your very own laboratory. Don't be daunted by the thought of expensive, high tech gear - with a little creativity and resourcefulness, you can create a well stocked workspace that would make even the most seasoned researchers green with envy.

Start by scouring your local thrift stores, garage sales, and online marketplaces for hidden gems. You'd be surprised at the gems you can uncover - from vintage microscopes to beakers, test tubes, and pipettes. Embrace the thrill of the hunt and let your inner MacGyver shine as you transform ordinary household items into essential scientific tools.

Of course, you'll also need to invest in some key items to ensure the success of your experiments. A reliable heat source, such as a hot plate or Bunsen burner, will be your best friend when it comes to temperature sensitive reactions. Ensure you have a sturdy work surface, like a repurposed kitchen table or a dedicated lab bench, to create a stable foundation for your experiments.

Don't forget the importance of safety equipment, such as goggles, gloves, and a lab coat. These not only protect you from potential hazards but also add an air of authenticity to your backyard lab. And while we're on the topic of safety, make sure to familiarize yourself with proper laboratory protocols and establish a designated storage area for any hazardous materials.

With your equipment and supplies in place, you're well on your way to building the ultimate backyard biotechnology haven. But the real excitement lies in the experiments themselves, so let's dive into that next.

Conducting Experiments at Home

Now that you've gathered your equipment and supplies, it's time to put them to the test and dive into the world of hands on biotechnology experiments. But before you start mixing chemicals and pipetting DNA, it's vital to have a solid plan in place.

Begin by identifying your areas of interest within the vast field of biotechnology. Are you fascinated by genetic engineering and want to explore DNA manipulation? Or perhaps you're more intrigued by the wonders of microbiology and the hidden universe of microorganisms. Whatever your passion, let it guide your experimental journey.

Once you've narrowed down your focus, it's time to start brainstorming experiment ideas. Scour the internet, read scientific journals, and seek inspiration from the pioneers in your chosen field. Don't be afraid to think outside the box - some of the most original discoveries have come from unexpected places.

As you investigate into your experiments, remember to document every step with meticulous care. Keep detailed records of your observations, measurements, and the results of your trials. This not only helps you track your progress but also lays the foundation for potential future collaborations or publications.

And speaking of collaborations, don't be afraid to reach out to like minded individuals online or in your local community. Joining forces with fellow backyard scientists can open up a world of new possibilities, from shared resources to collective problem solving. Who knows, you might even stumble upon the next scientific breakthrough together.

Remember, the beauty of a backyard lab lies in the freedom to explore, experiment, and push the boundaries of what's possible. Embrace the thrill of discovery, and let your curiosity be your guide as you start on this exciting biotechnology adventure.

Mastering Lab Techniques and Safety

As you dive deeper into your backyard biotechnology explorations, it's essential to master the fundamental lab techniques and, most importantly, prioritize safety. After all, working with living organisms and potentially hazardous materials requires a keen eye for detail and a steadfast commitment to responsible practices.

Start by familiarizing yourself with the basics of aseptic technique – the art of maintaining a sterile environment to prevent contamination. Proper handwashing, the use of sterile equipment, and the establishment of designated "clean" zones will become your mantra as you navigate the delicate world of microbiology.

Mastering pipetting is another essential skill that will serve you well in your backyard lab. Precise liquid handling is the backbone of countless experiments, from DNA extraction to cell culture. Practice makes perfect, so don't be afraid to start small and gradually increase the difficulty of your pipetting tasks.

Laboratory safety should be your top priority, and that means familiarizing yourself with the proper handling and storage of chemicals, as well as the appropriate personal protective equipment (PPE) for your experiments. Invest in high quality goggles, gloves, and a lab coat to safeguard yourself from potential hazards.

But safety extends beyond just physical protection – it's also about maintaining a organized, well maintained workspace. Establish clear protocols for waste disposal, ensure proper ventilation, and keep a well stocked first aid kit on hand. These simple precautions can make all the difference in the event of an unexpected incident.

As you refine your lab techniques and safety practices, you'll find that your confidence and efficiency will grow exponentially. With a solid foundation in these critical skills, you'll be well on your way to becoming a true backyard biotechnology master – one experiment at a time.

Networking and Mentorship

Connecting with Professionals in the Field

Building a vigorous network in the biotechnology industry is essential for your success, but it's not as simple as spamming endless LinkedIn connection requests. No, my friend, you need to get out there and make real, genuine connections – the kind that will make people remember you as more than just another face in the crowd.

Start by attending local industry events, conferences, and meetups. These are goldmines for making valuable connections, but don't just stand in the corner nursing a lukewarm cup of coffee. Mingle, strike up conversations, and be the one people remember as the quirky yet charismatic individual who had that, uh, interesting theory about how genetically modified kombucha could solve the global energy crisis.

And for the love of all things biotech, put down your phone and step away from the screen. Real connections happen face-to-face, so be present, make eye contact, and actually listen to what people are saying. You never know where a random chat about your shared passion for fermentation could lead – maybe to a game changing collaboration, or at the very least, a lifetime supply of deliciously fizzy beverages.

Seeking Out Experienced Mentors

As you navigate the winding path of biotechnology research, finding an experienced mentor can be the difference between a smooth journey and a chaotic stumble through the lab equipment. These seasoned professionals can offer indispensable guidance, wisdom, and a shoulder to lean on when the going gets tough.

But where do you even begin? Start by identifying researchers, industry leaders, or professors whose work inspires you. Reach out to them, not with a generic "I want you to be my mentor" message, but with a thoughtful, personalized proposal. Highlight your shared interests, explain why you admire their work, and offer to bring something unique to the table – maybe you've got a secret talent for brewing the perfect cup of microbially improved coffee, or you just happen to know the perfect spreadsheet formula for tracking experimental data.

Don't be discouraged if your initial outreach falls on deaf ears. Mentorship is a two way street, and it can take time to find the right fit. Keep putting yourself out there, attend events where your potential mentors might be, and be patient. The right mentor will recognize your passion and potential, and they'll be eager to help you navigate the complex and ever changing world of biotechnology.

Leveraging Internships and Volunteer Opportunities

As you begin on your biotechnology journey, don't just focus on the fancy degree programs and research grants – sometimes, the most valuable experiences come from getting your hands dirty in the lab, or volunteering your time and skills to support the greater good.

Internships, whether at a local biotech startup or a renowned research institution, can provide you with hands on experience, exposure to cutting edge technologies, and the chance to build lasting connections. And let's be real, who doesn't love the idea of getting to wear a lab coat and safety goggles on a regular basis? It's like playing dress up, but with a much cooler (and more scientifically significant) wardrobe.

But internships aren't the only way to gain valuable experience. Look for volunteer opportunities at non profit organizations, community labs, or even your university's research facilities. Not only will you be contributing to important work, but you'll also have the chance to network with a diverse array of professionals, from budding researchers to seasoned biotech veterans. And who knows, maybe your unpaid labor will lead to a life changing introduction or a sudden stroke of inspiration that propels you towards your next big breakthrough.

Advancing Your Skills: Hands on Experience

Participating in Research Projects

Dive headfirst into the exhilarating world of biotechnology research! As a budding scientist, there's no better way to hone your skills and expand your knowledge than by rolling up your sleeves and getting your hands dirty. Whether it's collaborating with a team of researchers on a cutting edge project or spearheading your own independent study, the opportunities for hands on experience are endless.

Start by scouring the internet, networking with professionals, and scouting out local universities or research institutions that offer exciting research opportunities for aspiring scientists. Don't be afraid to think outside the box – some of the most trailblazing discoveries have been made by those who dared to venture off the beaten path.

Once you've found a project that piques your interest, be prepared to bring your A game. Embrace the steep learning curve, ask questions until your brain hurts, and don't be afraid to make the occasional blunder. After all, that's how we learn and grow, right? Roll up your sleeves, dive into the data, and let your inner mad scientist shine!

Attending Workshops and Conferences

Forget the stuffy, lecture based conferences of the past – the biotech world is all about getting down and dirty with the latest innovations and trends. Attending industry leading workshops and conferences is a surefire way to supercharge your biotechnology knowledge and network like a pro.

Imagine being the first to hear about original advancements in gene editing, rubbing elbows with the pioneers of synthetic biology, or diving deep into the mind bending world of biomimicry. These events are where the magic happens, and you'll have a front row seat to the action.

But it's not just about soaking up information – these gatherings are also prime opportunities to showcase your own work, pitch your latest brainchild, and make essential connections that could propel your career to new heights. So, get ready to channel your inner extrovert, practice your elevator pitch, and get ready to make a lasting impression on the movers and shakers of the biotech world.

Honing Your Critical Thinking and Problem Solving Abilities

In the ever evolving terrain of biotechnology, the ability to think critically and solve complex problems is the ultimate superpower. Sure, you may have a solid foundation in the sciences, but to truly thrive in this field, you'll need to develop a sharp, analytical mindset that can tackle even the most perplexing challenges.

Start by embracing the art of questioning everything. Don't just accept the status quo – question it, poke holes in it, and explore alternative perspectives. Develop a insatiable curiosity that drives you to dig deeper, experiment with new approaches, and push the boundaries of what's possible.

Sharpen your problem solving skills by immersing yourself in a wide range of hands on projects and simulations. Tackle complex case studies, engage in lively debates, and don't be afraid to get your hands dirty with some good old fashioned trial and error. The more you challenge yourself, the more you'll develop the resilience, creativity, and adaptability needed to succeed in this fast paced, ever evolving field.

Navigating the Academic World

Applying to Graduate Programs

Initiating on a graduate program in biotechnology can be a thrilling and revolutionary experience, but the application process can also feel daunting. Fear not, my fellow science enthusiasts! With a little strategic planning and a healthy dose of determination, you can navigate the academic scene with confidence.

When it comes to choosing the right graduate program, don't just go for the shiny, well known names - dig deeper. Scour the university websites, explore research areas that pique your interest, and reach out to current students. They can offer indispensable realizations into the program's culture, mentorship opportunities, and whether it harmonizes with your unique career aspirations.

Now, let's talk about crafting that killer application. Forget the one size-fits all approach - each program has its own quirks and preferences. Pay close attention to the specific requirements, and tailor your statement of purpose accordingly. Weave in your personal story, your passion for the field, and how your experiences have shaped your scientific journey. And don't be afraid to let your personality shine through - admissions committees are looking for more than just a list of credentials.

But the application process doesn't end there. Securing strong letters of recommendation can make all the difference. Reach out to professors, research mentors, or industry professionals who can speak to your skills, work ethic, and potential for success. Remember, these aren't just letters - they're your personal cheerleaders, vouching for your awesomeness.

Finally, when it comes to financing your graduate education, explore every avenue. Research assistantships, teaching fellowships, and external scholarships can be game changers. Don't be afraid to get creative - maybe you've got a secret talent for juggling flaming beakers, and you can parlay that into a gig as the department's resident entertainment specialist. (Just make sure to clear that with the safety committee first.)

Excelling in Research and Coursework

You've conquered the application process and landed a spot in your dream graduate program. Congratulations, you ambitious biotechnology rockstar! Now, it's time to dive headfirst into the world of research and coursework, where the real magic happens.

As you navigate the academic situation, remember to approach every challenge with a curious and open mind. Embrace the opportunity to explore new frontiers, collaborate with brilliant minds, and push the boundaries of scientific understanding. Trust me, the thrill of discovering something entirely new will keep you fueled, even on those late nights in the lab.

But it's not all about the lab coat and pipette - excelling in your coursework is equally important. Treat your classes like interactive puzzles, where each lecture, discussion, and assignment is a chance to hone your critical thinking skills. Don't be afraid to ask questions, challenge assumptions, and engage in lively debates. After all, the best scientists are often the ones who aren't afraid to color outside the lines.

And speaking of thinking outside the box, don't limit yourself to the curriculum. Seek out interdisciplinary opportunities, attend guest lectures, and immerse yourself in the rich intellectual community that surrounds you. You never know when a chance encounter with a quantum physicist or a marine biologist might inspire your next trailblazing discovery.

Remember, the academic world is a delicate balance of dedication, curiosity, and the occasional strategic caffeine boost. Embrace the highs and the lows, learn from your mistakes, and never lose sight of the big picture. Because when you've got a passion for biotechnology that burns brighter than a supernova, the possibilities are truly endless.

Balancing Academia and Personal Life

As you dive headfirst into the world of graduate research and coursework, it's easy to become consumed by the persistent pursuit of scientific excellence. But let me let you in on a little secret: the true key to success lies in finding a harmonious balance between your academic aspirations and your personal well being.

First and foremost, let's address the elephant in the room: the dreaded "imposter syndrome." It's a common affliction that can plague even the most brilliant minds, convincing you that you're somehow not good enough or that you don't belong. Well, I'm here to tell you that it's all a load of hooey. You earned your place in this program, and you've got the passion and skills to thrive. So, embrace your inner mad scientist and let that confidence shine through.

Now, about that work life balance. It's easy to get so caught up in the world of academia that you forget to take care of yourself. But trust me, your research and your mental health go hand in-hand. Make time for self care, whether it's a weekly dance party in the lab, a rejuvenating hike in the great outdoors, or simply carving out time for a good Netflix binge session. Your brain and your body will thank you.

And let's not forget the importance of promoting a supportive network. Surround yourself with like minded individuals who understand the unique challenges of graduate life. Join student organizations, attend social events, and don't be afraid to lean on your peers for support. After all, who better to commiserate with over the trials and tribulations of experimental design than your fellow biotechnology enthusiasts?

Remember, your academic pursuits are just one piece of the puzzle. Keep your eyes on the prize, but don't forget to savor the journey. Embrace the occasional detour, laugh at the inevitable setbacks, and always remember to celebrate your victories, no matter how small. Because at the end of the day, this is your story to write, and it's going to be one heck of a page turner.

Embracing the Entrepreneurial Spirit

Identifying Market Opportunities

Ah, the entrepreneurial spirit - it's like a wild, untamed beast that lurks within each of us, just waiting to be let loose. But before you go chasing your biotech dreams, you need to do a little bit of reconnaissance. Because let's face it, the world of business is a jungle, and the only way to survive is to be one step ahead of the competition.

Start by keeping your eyes peeled for emerging trends and unmet needs in the biotechnology industry. Scour the news, attend industry events, and strike up conversations with fellow scientists and innovators. You never know where the next big idea might come from - it could be a chance encounter at the coffee machine or a random daydream about genetically engineered unicorns. (Don't laugh, stranger things have happened!)

Once you've identified a juicy opportunity, it's time to dig deeper. Analyze the market, study your potential customers, and assess the competition. But don't just crunch the numbers - get out there and talk to real people. Engage with your target audience, understand their is a difficulty, and find out what keeps them up at night. This is where your backyard experiments and networking skills will really come in handy.

Remember, the key to success isn't just about finding a good idea - it's about finding the *right* idea. The one that coordinates with your unique strengths, passions, and expertise. Don't be afraid to think outside the Petri dish and explore unconventional solutions. After all, the most original innovations often come from the most unexpected places.

Developing Original Products or Services

Alright, you've done your homework and identified a promising market opportunity. Now it's time to put on your mad scientist hat and get to work on that novel product or service. But beware, this isn't your typical lab experiment - you're about to enter the wild world of product development, where the only constant is change.

Start by embracing your inner MacGyver. Scrounge up whatever resources you can get your hands on, whether it's that dusty old centrifuge in your backyard lab or a team of eager interns. Get creative, think outside the box, and don't be afraid to take a few risks. After all, the greatest breakthroughs often come from the most unconventional approaches.

As you start to bring your creation to life, remember to stay flexible and adaptable. The biotechnology scene is ever evolving, and what might have been a game changing idea one day could be obsolete the next. So be prepared to shift, iterate, and constantly refine your product or service. Listen to your customers, gather feedback, and be willing to make tough decisions – even if it means scrapping your beloved project and starting from scratch.

And don't forget to protect your intellectual property! As your biotech business grows, you'll need to navigate the complex world of patents, trademarks, and copyrights. Consult with a savvy legal team to ensure your innovations are properly safeguarded, and don't be afraid to get a little scrappy when it comes to defending your turf.

Crafting a Successful Business Plan

You've got the passion, the ideas, and the drive – now it's time to put it all together into a killer business plan. But don't let the mere mention of the word "plan" send you running for the hills. Think of it as a roadmap for your biotech startup, a guiding light that will help you navigate the often turbulent waters of entrepreneurship.

Start by painting a clear picture of your vision – where do you see your biotech venture in 5, 10, or even 20 years? Get specific, and don't be afraid to dream big. After all, the most successful entrepreneurs are the ones who can see the world as it could be, not just as it is.

Next, dive into the nitty gritty details of your business model. What products or services will you offer? Who are your target customers, and how will you reach them? How will you price your selections, and how will you generate revenue? These are the kinds of questions you'll need to answer with razor sharp precision. And don't forget to factor in the all important financial projections – after all, even the most brilliant biotech idea is useless if it can't turn a profit.

But crafting a successful business plan isn't just about crunching numbers and forecasting growth. It's also about telling a compelling story – one that will captivate investors, partners, and potential customers alike. Weave in your personal narrative, your unique approach to problem solving, and your unwavering commitment to making a difference in the world of biotechnology. After all, sometimes the difference between a good idea and a great one comes down to sheer passion and storytelling prowess.

So don't be daunted by the prospect of creating a business plan – embrace it as an opportunity to articulate your vision, refine your strategy, and position your biotech startup for success. Who knows, with a little bit of elbow grease and a whole lot of moxie, you just might end up being the next rockstar of the biotech world.

Funding Your Biotechnology Endeavors

Exploring Research Grants and Fellowships

Ah, the grand adventure of securing funding for your biotechnology dreams - it's like a high stakes game of financial Tetris, where you have to meticulously arrange your puzzle pieces (read: grant proposals) to land that life changing jackpot. But fear not, my fellow mad scientists, for the world of research grants and fellowships is a veritable goldmine, waiting to be revealed by those with the tenacity and strategic finesse to navigate its labyrinthine corridors.

First and foremost, embrace your inner investigative reporter. Scour the internet, comb through academic databases, and network like a seasoned politician on the campaign trail. Uncover every nook and cranny where funding opportunities might be lurking, from government agencies to private foundations, each with their own unique set of requirements and priorities. It's like a high stakes scavenger hunt, where the prize is the financial freedom to turn your mad science dreams into reality.

But don't just blindly apply to every grant you can find - that's a surefire way to end up with a stack of rejection letters taller than your backyard lab equipment. No, my friends, the key is to tailor your proposals with the precision of a surgeon wielding a scalpel. Research the funding agency's mission, understand their specific areas of interest, and then craft a pitch that speaks directly to their needs and aspirations. It's all about making them feel like you're the perfect match, the yin to their yang, the peanut butter to their jelly.

And when that glorious day comes, and the funding gods smile upon you, resist the urge to do a victory dance on your laboratory bench. Remember, the real work is just beginning. Steward those precious funds with the care and diligence of a parent guarding their child's first steps, ensuring every penny is put to good use in the pursuit of your biotechnological vision. After all, the world is watching, and you've got a legacy to build.

Crowdsourcing and Crowdfunding Strategies

In the high stakes world of biotechnology, sometimes it takes a village to raise a lab coat wearing child. Enter the realm of crowdsourcing and crowdfunding, where the collective power of the masses can be used to bring your scientific dreams to life. It's like a modern day scientific barn raising, where everyone chips in a few bucks to help you erect the towering edifice of your next breakthrough.

But don't be fooled - crowdfunding isn't just about posting a cute video and waiting for the donations to roll in. Oh no, my friends, it requires a strategic masterplan, a carefully curated campaign that captivates the hearts and minds of potential backers. You've got to be part showman, part storyteller, weaving a narrative that makes people feel like they're not just investing in a project, but in a movement, a revolution in the making.

Start by crafting a compelling and visually striking pitch that speaks directly to the passions and interests of your target audience. Maybe it's a cutting edge technique to grow biofuel producing algae in your backyard, or a plan to engineer a new strain of bacteria that can clean up toxic waste. Whatever it is, make sure it's bold, original, and above all, undeniably captivating.

But the real secret sauce? Engagement, engagement, engagement. Don't just sit back and wait for the money to roll in - get out there and grow a loyal following, a community of fellow science enthusiasts who will support your cause and spread the word like digital age town criers. Share updates, hold live Q&As, and give your backers a behind the-scenes look at the mad science in action. After all, they're not just investors - they're partners in your quest to change the world, one test tube at a time.

Navigating the World of Venture Capital

Ah, the siren call of venture capital - where the deep pocketed and the daring come together to forge the future, one high risk, high reward gamble at a time. If you're a biotechnology entrepreneur with an insatiable appetite for adventure and a penchant for the unconventional, then lend me your ear, my friend, for I have a tale or two to share about the treacherous, yet exhilarating journey of securing those elusive VC dollars.

First and foremost, forget about the stuffy boardroom presentations and the buttoned up PowerPoint slides. In the world of venture capital, it's all about the wow factor, the ability to captivate your audience with a vision so bold, so audacious, that they can't help but be swept up in the sheer magnitude of your ambition. Think big, think disruptive, think "I'm going to change the world, and I need your money to do it."

But don't mistake that bravado for a free pass - these venture capitalists are savvy, seasoned sharks, and they'll smell your fear and inexperience from a mile away. You've got to come armed with a rock solid business plan, a deep understanding of your market, and the kind of infectious enthusiasm that makes them believe you're the one to lead the charge. It's like a high stakes game of intellectual chess, where every move you make has to be calculated, every word you utter must be dripping with confidence and conviction.

And once you've secured that golden ticket, the real work begins. Be prepared to navigate a complex web of expectations, milestones, and reporting requirements – it's like juggling a dozen test tubes filled with volatile chemicals, all while performing a tightrope act over a pit of ravenous venture capital piranhas. But hey, if you can survive that, you'll be well on your way to turning your biotechnology dreams into a reality, with the backing of some of the most influential movers and shakers in the game.

Overcoming Challenges and Setbacks

Developing Resilience and Perseverance

Ah, the joys of being a budding biotechnology pioneer - the endless nights spent hunched over a microscope, the frustration of failed experiments, the constant battle against the forces of bureaucracy and red tape. Welcome to the wild ride that is the life of a biotechnology researcher! But fear not, my fellow scientists, for within the chaos lies the opportunity to forge a path of true greatness.

Resilience, my friends, is the backbone of any successful biotechnology journey. When your carefully constructed experiment crumbles before your eyes, or when that grant application gets rejected for the third time, it's easy to want to throw in the towel and call it quits. But that's where the magic happens - in the moments where you stare adversity in the face and say, "Not today, my friend. Not today."

Nurture a mindset of unwavering determination, a tireless pursuit of knowledge, and an unshakable belief in your own abilities. Failure is not the end, but rather a chance to learn, to grow, to refine your approach. Embrace those setbacks as stepping stones to your ultimate triumph. Channel your inner Rocky Balboa and keep punching, even when the odds seem stacked against you.

Persistence, too, is a vital weapon in your arsenal. Breakthroughs in biotechnology don't happen overnight; they are the result of painstaking, meticulous work, day in and day out. Nurture the discipline to stay the course, to keep chipping away at the problem, even when the light at the end of the tunnel seems so far away. Trust the process, trust your instincts, and trust in the power of your own indomitable spirit.

Learning from Failures and Mistakes

Ah, the joys of scientific discovery - where failure is not just a possibility, but an inevitability. Embrace it, my friends, for it is in the depths of our mistakes that we find the greatest opportunities for growth and innovation.

Every time that carefully crafted experiment falls flat on its face, or that hypothesis you were so sure about gets completely debunked, it's a chance to learn. Peel back the layers of your failure, dissect it with a critical eye, and uncover the valuable lessons hidden within. What went wrong? What could you have done differently? How can you apply this newfound knowledge to your next foray into the unknown?

Failures are the stepping stones to success, the fuel that propels us forward. Embrace them, celebrate them, even. Because the day you stop making mistakes is the day you stop learning, the day your scientific curiosity and thirst for discovery begins to wither away.

Remember, the greatest minds in biotechnology history didn't reach the pinnacles of their success without a few (or a few hundred) stumbles along the way. Thomas Edison, the legendary inventor, famously said, "I have not failed. I've just found 10,000 ways that won't work." Adopt that mindset, my fellow researchers, and watch as your failures transform into the building blocks of your future triumphs.

Maintaining a Positive Mindset

In the high stakes, high pressure world of biotechnology research, it's easy to get bogged down by the weight of it all. The failed experiments, the endless grant applications, the bureaucratic red tape – it can feel like an endless barrage of obstacles standing in the way of your dreams. But, my friends, it is in these moments of adversity that your true mettle is tested.

Nurture a mindset of unwavering positivity, a tireless optimism that refuses to be extinguished, no matter how dim the circumstances may seem. When the going gets tough, and the path forward appears shrouded in darkness, remember the reason you started on this journey in the first place – the boundless wonder of the living world, the thrill of untangling life's most elaborate mysteries, the opportunity to create real, tangible change that can improve the lives of countless individuals.

Surround yourself with a support network of like minded individuals, people who will lift you up when the weight of the world threatens to pull you down. Lean on your mentors, your peers, your fellow biotechnology pioneers, and draw strength from their experiences, their triumphs, and their unwavering belief in your potential.

And when the self doubt creeps in, when the imposter syndrome starts to whisper poisonous lies in your ear, remember this: you are a trailblazer, a visionary, a scientist with the power to shape the future of our world. Embrace that truth, hold it close to your heart, and let it fuel your every step forward. For in the end, it is not the challenges we face, but the indomitable spirit with which we confront them, that will define our legacy as biotechnology pioneers.

Navigating the Regulatory Terrain

Understanding Biotechnology Laws and Regulations

Welcome to the wild world of biotechnology regulations! It's a labyrinth of acronyms, legalese, and enough red tape to make your head spin. But fear not, my aspiring biotech pioneer, for we shall navigate these treacherous waters together. Buckle up, because it's about to get interesting.

First and foremost, let's talk about the alphabet soup of governing bodies that oversee the biotech industry. You've got the FDA, the EPA, the USDA, and a whole host of other agencies that have their fingers in the pie. It's like a game of regulatory Tetris, where you've got to fit all the pieces together just right to stay on the right side of the law.

But don't let the bureaucracy intimidate you. Embrace it, my friends! Dive headfirst into those regulatory guidelines and use them to your advantage. After all, who knows the rules better than the ones who write them? Become a regulatory ninja, mastering the art of navigating the system and turning it to your favor.

Of course, it's not all about the legalities. Ethical considerations are just as critical in the world of biotechnology. You'll need to navigate the murky waters of bioethics, where there are as many opinions as there are strands of DNA. But fear not, for with a strong moral compass and a commitment to responsible research, you can chart a course that keeps you on the right side of history.

Ensuring Ethical and Responsible Practices

Ah, the delicate dance of ethics and biotechnology. It's a minefield of complex issues, where the line between progress and risk is often blurred. But as a budding biotech pioneer, it's your job to be the guardian of moral integrity, the keeper of the ethical torch.

First and foremost, you'll need to develop a keen understanding of the various ethical frameworks at play. From the principles of informed consent to the debates surrounding gene editing, you'll need to be well versed in the philosophical and practical implications of your work.

But it's not enough to simply know the rules – you've got to live them. Nurture a culture of ethical decision making within your lab or organization. Encourage open dialogue, promote a spirit of transparency, and never compromise your principles in the pursuit of progress.

And remember, ethical considerations aren't just about what you do in the lab – they extend to how you interact with the broader community. Be a ambassador for responsible biotechnology, educating the public, collaborating with stakeholders, and always keeping the greater good in mind.

It's a delicate balance, to be sure, but one that separates the true pioneers from the also rans. So embrace the challenge, my friends, and let your ethical compass be your guide through the uncharted waters of the biotech setting.

Collaborating with Regulatory Agencies

Ah, the dance with the regulatory powers that-be – it's a tango of epic proportions, my friends. But fear not, for with a little finesse and a whole lot of patience, you can master the steps and emerge victorious.

First and foremost, it's essential to develop a deep understanding of the various agencies that hold sway over the biotech industry. From the FDA's stringent drug approval process to the EPA's watchful eye on environmental impact, you'll need to be intimately familiar with the key players and their respective jurisdictions.

But knowledge is only half the battle. The real secret lies in cultivating strong relationships with the regulatory agencies. Think of them as your dance partners – you need to be in perfect sync to pull off the moves with grace and precision.

Start by reaching out, introducing yourself, and establishing open lines of communication. Attend industry events, participate in public comment periods, and make yourself a visible and engaged member of the regulatory community. Show them that you're not just a faceless biotech entity, but a passionate advocate for responsible innovation.

And when it comes time to navigate the regulatory maze, approach it with a spirit of collaboration, not confrontation. Work closely with agency representatives, provide clear and comprehensive documentation, and be prepared to address any concerns they may have. Remember, they're not the enemy – they're gatekeepers tasked with ensuring the safety and integrity of the industry.

So, my fellow biotech pioneers, embrace the dance, master the steps, and let the regulatory tango be the foundation upon which you build your empire of scientific wonder and discovery.

Translating Research into Real world Applications

Identifying Practical Solutions to Global Challenges

As a biotechnology researcher, your work holds the potential to address some of the most pressing issues facing our world. From tackling hunger and disease to developing sustainable energy solutions, the innovations that emerge from your lab can have a significant impact on humanity. But how do you ensure your research translates into real world applications that truly make a difference?

The key lies in maintaining a laser sharp focus on identifying practical solutions to global challenges. Start by immersing yourself in the grand challenges of our time – whether it's combating climate change, improving access to clean water, or revolutionizing medical treatments. Familiarize yourself with the latest developments, trends, and is a problem in these areas, and let that knowledge guide your research agenda.

Collaborate with experts from diverse fields, from policy makers and social scientists to industry leaders and community stakeholders. By encouraging interdisciplinary partnerships, you can gain a deeper understanding of the real world implications and implementation barriers associated with your work. This cross pollination of perspectives will help you refine your research approach and ensure your findings have a tangible impact.

Don't be afraid to think outside the petri dish. Explore novel applications for your discoveries, even if they may seem unconventional or unrelated to your initial area of focus. Some of the most original innovations come from unexpected intersections of disciplines. Embrace your inner maverick and be willing to challenge the status quo.

Bridging the Gap Between Lab and Industry

Translating your research from the lab to the real world can be a daunting task, but it's a critical step in ensuring your work has a meaningful impact. Building strong connections with industry partners can help you navigate this transition and make accessible new avenues for collaboration and commercialization.

Start by immersing yourself in the business situation – attend industry conferences, network with professionals, and stay up-to-date on market trends and is a challenge. Understand the unique challenges and constraints that companies face, and identify how your research could provide original solutions to their problems.

Develop a strategic mindset and learn to speak the language of the business world. Craft a compelling narrative that showcases the practical applications and commercial potential of your work. Embrace the art of pitching and be prepared to demonstrate the value proposition of your research in a way that relates with industry stakeholders.

Explore avenues for collaborative research, licensing agreements, or even starting your own venture. Don't be afraid to take on an entrepreneurial mindset and explore the possibility of commercializing your discoveries. With the right partnerships and support, you can bridge the gap between the lab and the marketplace, and see your innovations come to life in the real world.

Communicating Complex Concepts to Diverse Audiences

As a biotechnology pioneer, you possess a wealth of specialized knowledge and technical expertise. But to truly drive change and impact the world, you must master the art of translating your complex scientific concepts into accessible, engaging, and significant communication.

Adapt your communication style to the specific needs and backgrounds of your audience. Whether you're presenting to a room full of policymakers, pitching to investors, or educating the general public, your ability to break down complex ideas and convey their significance will be critical.

Embrace the power of storytelling – weave narratives that capture the human element and the real world implications of your research. Use vivid analogies, relatable examples, and a touch of humor to bring your work to life and make it relate with your audience.

Use multimedia tools and visual aids to improve your communication. Infographics, interactive demonstrations, and captivating videos can help simplify and increase your message. Remember, the way you present your ideas can be just as important as the content itself.

Grow a genuine passion for sharing your knowledge and inspiring others. Your enthusiasm and commitment to making a difference will shine through and captivate your audience, whether they're seasoned experts or curious bystanders. By communicating with clarity, creativity, and conviction, you can bridge the gap between the technical and the accessible, and equip others to join you in shaping the future of biotechnology.

Embracing Diversity and Inclusion

Promoting Inclusive Environments

In the dynamic and ever evolving world of biotechnology, nurturing inclusive environments is not just a moral imperative, but a strategic necessity. After all, the most original discoveries often arise at the intersection of diverse perspectives and experiences. So, forget the stuffy boardrooms and the ivory tower mentality - it's time to roll up our sleeves and create workspaces that celebrate the unique talents and backgrounds of all their inhabitants.

Start by banishing the "brogrammer" culture and its toxic undertones. Sure, we all love a good lab dance off, but let's make sure the music selection represents the eclectic tastes of the whole team. Heck, why not invite the local mariachi band to liven up the holiday party? And while we're at it, let's ditch the mandatory "beer pong Fridays" in favor of something a little more inclusive, like a potluck where everyone brings a dish that reflects their cultural heritage.

But diversity is more than just good vibes and cultural showcases. It's about actively dismantling the widespread barriers that have long kept marginalized groups from reaching their full potential. So, let's put our money where our mouth is and invest in mentorship programs, scholarships, and internships that target underrepresented communities. And when it comes to hiring, let's challenge ourselves to look beyond the traditional résumé filters and focus on the unique skills and perspectives that each candidate brings to the table.

Remember, true inclusion isn't just about checking boxes or ticking off diversity quotas. It's about creating an environment where everyone feels allowed to bring their authentic selves to work, where they can take risks, ask questions, and challenge the status quo without fear of retribution. It's about celebrating our differences and channeling them into new solutions that push the boundaries of what's possible in the world of biotechnology.

Enabling Underrepresented Groups in Biotechnology

In the high stakes world of biotechnology, the playing field has far too often been tilted in favor of those with the "right" pedigree - the Ivy League graduates, the well connected scions of industry titans, the prodigies who've been groomed for success since birth. But what about the rest of us, the mavericks and the misfits, the passionate individuals who may not have had the same advantages, but whose unique perspectives and lived experiences could be the key to opening up the next breakthrough?

It's time to shatter the glass ceiling and enable the underrepresented voices in our field. Let's start by increasing the stories of those who have defied the odds and carved out their own paths, whether it's the first generation college student who fought tooth and nail to secure a coveted research fellowship or the single parent who juggles family responsibilities while pursuing their PhD.

But we can't stop there. We need to actively invest in programs and initiatives that provide access to the resources, mentorship, and support that these trailblazers need to thrive. Think mentorship circles, coding bootcamps for aspiring bioinformaticians, or even a "biotechnology boot camp" that demystifies the inner workings of the industry and equips participants with the skills and confidence to navigate it.

And let's not forget about the power of role models. By showcasing the diverse array of scientists, entrepreneurs, and is a trendsetter who are shaping the future of biotechnology, we can inspire the next generation to dream bigger and reach higher. Invite them to speak at your conferences, feature them in your marketing materials, and make them an fundamental part of your organization's fabric.

Remember, allowing underrepresented groups in biotechnology isn't just the right thing to do - it's a strategic imperative. The more we embrace the wealth of untapped talent and innovation that exists beyond the traditional circles, the more we position ourselves to tackle the complex challenges that lie ahead. So, let's roll up our sleeves and get to work, because the future of our field depends on it.

Advocating for Equity and Representation

In the fast paced, high stakes world of biotechnology, it's easy to get caught up in the pursuit of scientific breakthroughs and the race to commercialize the latest cutting edge innovations. But as we strive to push the boundaries of what's possible, we must never lose sight of the fundamental principles of equity and representation that underpin true progress.

After all, what good are our technological advancements if they only benefit a select few? What happens when the life saving therapies we develop fail to reach the very communities that need them the most? That's why it's essential for us, as biotechnology pioneers, to be tireless advocates for a more inclusive and equitable future.

It starts with recognizing the deep-rooted biases and barriers that have long kept marginalized groups on the sidelines. We must confront the uncomfortable truth that our industry has often been a bastion of privilege, where access to education, funding, and professional networks has been heavily skewed towards those who already hold positions of power.

But rather than wring our hands and lament the challenges, let's channel that energy into concrete action. Let's use our influence and platforms to back policies and initiatives that level the playing field, whether it's pushing for more inclusive clinical trials, advocating for increased funding and research opportunities in underserved communities, or demanding greater representation on corporate boards and in leadership roles.

And let's not forget the power of our own voices. By speaking out against injustice, by intensifying the stories of those who have overcome adversity, and by using our platforms to shine a spotlight on the remarkable contributions of underrepresented individuals, we can inspire the next generation of biotechnology pioneers to dream bigger and reach higher.

Because at the end of the day, the future of our industry doesn't just depend on the latest technological breakthroughs - it hinges on our ability to create a more equitable and inclusive network that enables everyone to participate and thrive. So, let's roll up our sleeves and get to work, shall we? The world is waiting for the innovations that can only emerge when we embrace the full spectrum of human talent and potential.

Achieving Work Life Balance

Prioritizing Self care and Well being

In the constant pursuit of scientific excellence, it's easy to become consumed by the demands of your work. But let's be real, folks - you can't pour from an empty cup. If you don't take care of yourself, you'll be no good to anyone, let alone your original research. So, let's talk about how to keep that tank topped up and that brain firing on all cylinders.

First things first, ditch the savior complex. You're not a superhero, and you don't have to do it all. Learn to say "no" without guilt - your time and energy are precious, and you need to protect them fiercely. Stop burning the midnight oil and instead, establish a solid sleep routine. Trust me, those eureka moments come much easier when your synapses aren't firing on fumes.

And let's not forget the importance of good old fashioned exercise. You don't have to be a CrossFit advocate for, but get that body moving, even if it's just a brisk walk around the block. Not only will it keep your physique in tip top shape, but it'll also do wonders for your mental health. Put on your best '80s montage music and get that heart pumping – you'll be back in the lab, tackling that pesky experiment with renewed vigor.

Don't forget to refuel with wholesome, nourishing grub, too. We know you're tempted to mainline caffeine and energy drinks, but trust us, your body will thank you for a balanced diet. Fuel up on brain boosting superfoods, stay hydrated, and for the love of all things fermented, remember to take breaks and actually eat your meals.

And let's not forget the power of unplugging. In this hyper connected world, it's easy to get caught up in the endless scroll, but that's a surefire way to fry your circuits. Set boundaries, designate tech free zones and times, and learn to truly be present, whether it's with loved ones or just with your own thoughts. Your well being will thank you, and who knows, you might just stumble upon your next big idea in the shower.

Managing Time and Stress Effectively

Ah, the delicate dance of time management – the bane of every ambitious scientist's existence. Between grant proposals, lab experiments, and that pesky need for sleep, it's a wonder any of us ever get anything done. But fear not, my fellow biotechnology trailblazers, for there are ways to tame the ticking clock and keep your stress levels in check.

First, let's talk priorities. Grab a pen and paper (or, you know, open up a digital notepad – we're not cavemen) and do a brutal assessment of your to do list. What's truly essential, and what's just busy work masquerading as productivity? Be ruthless, my friend, and don't be afraid to delegate or even delete tasks that aren't moving the needle.

Next, implement a scheduling system that works for you. Whether it's the Pomodoro technique, time blocking, or good old fashioned calendar management, find a method that helps you stay focused and on track. And don't forget to schedule in breaks – those precious moments of mental respite will keep you sharp and ready to tackle the next challenge.

Speaking of challenges, let's address the elephant in the room: stress. It's the silent killer of productivity and creativity, and it's all too easy to let it creep in and take over. When the pressure's on, take a deep breath, step away from your work, and find healthy ways to decompress. Go for a walk, try meditation, or just blast some cheesy '90s pop and have a solo dance party – whatever works to reset your mind and body.

And remember, you're not alone in this. Reach out to your support network, whether it's colleagues, mentors, or loved ones. Talking through your struggles can do wonders for your mental well being, and who knows, you might even pick up a few time saving tips along the way.

Cultivating a Supportive Network

In the high stakes world of biotechnology, it's easy to feel like you're navigating a labyrinth of challenges all by your lonesome. But the truth is, you don't have to go it alone. Surround yourself with a network of like minded individuals who can lift you up, offer guidance, and remind you that you're not just a disembodied brain in a lab coat – you're a human being, and you matter.

Start by connecting with your peers – those fellow science geeks who understand the unique joys and tribulations of the research life. Join professional organizations, attend industry events, or even start your own virtual coffee klatch. These connections can provide a much needed sounding board, a shoulder to lean on, and maybe even a few raucous lab stories to bond over.

Don't forget to tap into the wisdom of experienced mentors, too. These seasoned veterans have been through the trenches and come out the other side, and they're often more than willing to share their hard earned realizations. Reach out, buy them a cup of coffee (or, let's be real, a craft beer), and pick their brains. You never know what nuggets of wisdom they might impart that could change the trajectory of your career.

And let's not forget the importance of maintaining a strong support system at home. Your loved ones – whether it's your partner, family, or closest friends – can be the rock that keeps you grounded when the tides of academic and professional life start to feel overwhelming. Lean on them, let them celebrate your victories, and don't be afraid to let them know when you're struggling. They're in your corner, and their unwavering support can make all the difference.

Remember, you're not just a scientist – you're a human being with complex needs and desires. By cultivating a diverse and supportive network, you're not only investing in your professional success, but also in your overall well being. So go forth, connect, and remember – you don't have to do this alone.

Leaving a Lasting Legacy

Mentoring the Next Generation of Biotechnology Pioneers

As you start on your journey as a biotechnology trailblazer, it's important to remember that your impact extends far beyond your own accomplishments. One of the most meaningful ways to leave a lasting legacy is by mentoring the next generation of scientists and innovators. Think of yourself as a guiding light, illuminating the pathways that future pioneers can follow.

Seek out opportunities to volunteer your time and expertise at local schools, community colleges, or university programs. Share your stories, your failures, and your triumphs. Inspire young minds by demonstrating the boundless possibilities that lie within the realm of biotechnology. Encourage them to ask questions, to challenge the status quo, and to never be content with the limits of what is currently known.

Offer internships or research opportunities within your own lab or organization. These hands on experiences can be revolutionary, allowing aspiring scientists to apply their classroom knowledge in a real world setting. Mentor them, not just as a superior, but as a collaborator and a fellow explorer. Impart the valuable lessons you've learned, from navigating the intricacies of experimental design to the art of effective communication.

Pay it forward by connecting your mentees with others in the field, expanding their professional network and opening doors to new opportunities. Celebrate their successes and offer guidance when they encounter obstacles. Your role as a mentor is not just to impart knowledge, but to grow the next generation of leaders who will push the boundaries of what's possible in biotechnology.

Contributing to the Advancement of the Field

As you progress in your biotechnology career, you have a unique opportunity to contribute to the advancement of the field in meaningful and influential ways. Whether it's through original research, novel product development, or shaping policy and regulatory frameworks, your contributions have the power to shape the future of biotechnology.

Engage actively in the scientific community by presenting your work at conferences, publishing in peer reviewed journals, and participating in collaborative research projects. These platforms not only showcase your expertise but also inspire and challenge your peers to push the boundaries of scientific discovery.

Consider taking on leadership roles within professional organizations, academic institutions, or industry associations. Use your voice to advocate for funding, resources, and policies that support the continued growth and development of biotechnology. Collaborate with policymakers, regulatory agencies, and industry leaders to ensure that the field evolves in an ethical, responsible, and sustainable manner.

Seek out opportunities to contribute to the development of new technologies, methods, or applications that address pressing global challenges. From tackling infectious diseases to developing sustainable energy solutions, your work can have a intense impact on the lives of people around the world. Embrace the responsibility that comes with being a biotechnology pioneer, and use your expertise to make a lasting difference.

Inspiring Others through your Achievements

As you navigate the ever evolving setting of biotechnology, your personal achievements and the stories you share have the power to inspire and authorize others. Your journey, with all its ups and downs, can serve as a demonstration to the revolutionary potential of passion, perseverance, and a constant pursuit of knowledge.

Be unapologetic in your celebration of your accomplishments, no matter how seemingly small they may be. Each milestone, from publishing your first research paper to securing a coveted grant, is a evidence to your hard work and dedication. Share these moments with the wider community, inspiring others to believe that their own dreams are within reach.

Use your platform to intensify the voices and experiences of underrepresented groups in the field of biotechnology. Shine a spotlight on the unsung heroes, the trailblazers who have paved the way, and the rising stars who are poised to redefine the future. By advocating for diversity and inclusion, you can inspire the next generation of biotechnology pioneers to believe that they, too, have a rightful place in this dynamic and rapidly evolving industry.

Embrace the power of storytelling to connect with a broader audience. Share the challenges you've overcome, the setbacks you've faced, and the lessons you've learned. Your vulnerability and authenticity can inspire others to embrace their own struggles and find the strength to persevere. By sharing your journey, you can ignite a spark within aspiring biotechnologists, equipping them to forge their own paths and leave an indelible mark on the world.

The Future of Biotechnology: Trends and Innovations

Emerging Technologies and their Implications

Hold onto your lab coats, folks, because the world of biotechnology is about to get a whole lot more exciting - and a little bit mind bending. As we peer into the crystal ball of scientific innovation, we're seeing a dazzling array of emerging technologies that are poised to transform the field. Brace yourselves for a wild ride, because the future is here, and it's more incredible than you could have ever imagined.

First up, let's talk about the rise of synthetic biology - the ultimate genetic Lego set for the 21st century. Imagine being able to design and engineer living organisms from scratch, like a mad scientist in a lab coat and goggles. With the power of CRISPR and other cutting edge gene editing tools, researchers are pushing the boundaries of what's possible, creating everything from bacteria that can devour plastic to yeast that can produce biofuels. The implications are staggering, from revolutionizing medicine to tackling global environmental challenges.

But that's just the tip of the iceberg. Wearable biotech is poised to become the new normal, with smart fabrics and implantable sensors that can monitor our health, deliver targeted drug treatments, and even improve our physical capabilities. Imagine a future where your shirt can detect the early signs of a heart attack or your contact lenses can provide real time data on your glucose levels. It's like having your very own personal medical lab strapped to your body 24/7.

And let's not forget about the rise of lab grown everything. From cultured meat to synthetic diamonds, the boundaries between the natural and the artificial are blurring. This "post nature" revolution is not only transforming industries but also challenging our very understanding of what it means to be human. As we learn to use the power of biology, the line between the living and the engineered becomes increasingly hazy.

But with great power comes great responsibility, and the emergence of these cutting edge biotechnologies also raises vital ethical and regulatory questions. How do we ensure that these innovations are used for the greater good, rather than to exploit or harm? How do we balance the boundless potential of these technologies with the need to protect individual privacy, safeguard the environment, and maintain the delicate balance of our living world? These are the challenges that will shape the future of biotechnology, as we navigate uncharted territory with both awe and trepidation.

Addressing Global Challenges with Biotechnology

In a world beset by daunting challenges – from climate change and food insecurity to the spread of deadly diseases – the field of biotechnology stands as a beacon of hope. As backyard scientists turned-lab pioneers, we have the unique opportunity to channel the power of living systems to tackle some of humanity's most pressing problems. So, strap on your lab coats and get ready to roll up your sleeves, because the future of the planet is in our hands.

Let's start with the big one: climate change. Biotechnology is poised to play a central role in our fight against global warming, from engineered microbes that can gobble up greenhouse gases to the development of sustainable biofuels and bioplastics. Imagine a world where our cars, our packaging, and even our clothing are made from renewable, biodegradable materials that don't contribute to the ever growing carbon footprint. It's a future that's within our grasp, thanks to the new minds in the world of biotechnology.

But it's not just about the environment – biotechnology is also a powerful tool in the battle against disease and food insecurity. With advancements in genomics, proteomics, and synthetic biology, we're opening up new ways to combat deadly pathogens, develop personalized medical treatments, and engineer crops that can thrive in even the harshest conditions. Imagine a world where a simple cheek swab can reveal your genetic predispositions, allowing you to take proactive steps to maintain your health, or where a genetically modified strain of rice can feed millions in drought stricken regions.

And the best part? We're just scratching the surface of what's possible. As the field of biotechnology continues to evolve, we're bound to uncover even more novel solutions to the global challenges that threaten our very existence. It's a future that's both exciting and terrifying, filled with both boundless potential and weighty responsibility. But for those of us who have caught the biotechnology bug, there's no turning back – we're in this for the long haul, determined to shape a better world, one experiment at a time.

Envisioning the Next Frontier of Scientific Discovery

Hold onto your lab goggles, folks, because the future of biotechnology is about to get downright cosmic. As we peer into the crystal ball of scientific innovation, we're seeing a dizzying array of breakthroughs that could propel us to the very edges of our understanding of the living world – and beyond.

First on the list? The quest to decipher the mysteries of the human genome. With the rapid advancements in DNA sequencing and gene editing technologies, we're on the cusp of a new era of personalized medicine, where our genetic code becomes the key to revealing our unique health needs and vulnerabilities. Imagine a future where a simple blood test can reveal the hidden predispositions lurking in your DNA, allowing you to take proactive steps to maintain your well being. It's a future that's both exciting and a little bit unsettling, as we grapple with the ethical implications of having such intimate knowledge of our own genetic makeup.

But the frontiers of biotechnology don't stop at our own species. As we explore deeper into the mysteries of the natural world, we're uncovering a dazzling array of life forms that could hold the key to solving some of humanity's most pressing problems. From extremophile bacteria that thrive in the most hostile environments to the detailed communication networks of plant communities, the diversity of life on our planet is a wealth of untapped potential. Imagine a future where we employ the power of these resilient organisms to clean up toxic waste, produce biofuels, or even terraform other planets – the possibilities are truly mind boggling.

And let's not forget about the thrilling prospect of venturing beyond our own world. As we continue to explore the cosmos, the question of extraterrestrial life becomes increasingly tantalizing. What kind of alien biology might we uncover? Could we one day establish a human presence on other planets, relying on the tools of biotechnology to create self sustaining is a system and thrive in the most inhospitable environments? The thought of it is enough to make any backyard scientist's heart race with excitement.

Yes, the future of biotechnology is a wild ride, filled with equal parts awe and trepidation. But for those of us who have caught the scientific bug, there's no turning back. We're on the cusp of a new frontier, where the boundaries between the natural and the artificial blur, and the very fabric of life is ours to manipulate and understand. So, let's strap on our lab coats, sharpen our pipettes, and get ready to chart the course for a future that's more incredible than we could have ever imagined.

Copyright 2024

Silas Meadowlark

www.ingramcontent.com/pod-product-compliance
Lightning Source LLC
Chambersburg PA
CBHW030505220526
45464CB00006B/2659